V2 *DAWN OF THE ROCKET AGE*

LEGEND FOR COVER PICTURE

1. Main alcohol supply
2. Main fuel turbopump
3. Compressed air flask
4. Blower for expansion system
5. Automatic alcohol valve
6. Insulation
7. Front linking ring
8. Integrated acceleration gauge and radios
9. Explosive charge
10. Fuze head
11. Fuze channel
12. Electric main fuze
13. Nitrogen flask
14. Plywood panel
15. Automatic circulator
16. Alcohol-water mixture
17. Alcohol line
18. Liquid oxygen (A-substance)
19. Insulated alcohol line
20. Alcohol mixing line
21. Fuel container (hydrogen peroxide)
22. Engine mount
23. Fuze container
    (Potassium permanganate)
24. Mixture distributor
25. Drive heater (powerplant)
26. Alcohol cooling lines
27. Electric motor
28. Electric hydraulics
29. Radio antenna
30. Jet rudder
31. Rudder control line
32. Air rudder

Joachim Engelmann

Schiffer Publishing Ltd

1469 Morstein Road, West Chester, Pennsylvania 19380

SOURCES

J. Engelmann, *Geheime Waffenschmiede Peenemünde*
German Museum, Munich (M)
Federal Archives, Koblenz (BA)
Engelmann Archives
Scheibert Archives

Translated from the German by Dr. Edward Force,
Central Connecticut State University.

Printed in the United States of America.
ISBN: 0-88740-233-X

This book originally published under the title,
V2 Aufbruch zur Raumfahrt,
by Podzun-Pallas Verlag, 6360 Friedberg 3 (Dorheim)
© 1985. ISBN: 3-7909-0241-1.

We are interested in hearing from authors with
book ideas on German military history.

Published by Schiffer Publishing, Ltd.
1469 Morstein Road
West Chester, Pennsylvania 19380
Please write for a free catalog.
This book may be purchased from the publisher.
Please include $2.00 postage.
Try your bookstore first.

Johannes Winkler, 1897-1946, pioneer of rocket technology.

The Development of the V2

Strictly speaking, the age of space rockets began at the moment when, in the spring of 1930, Wernher von Braun, and in the summer, then-captain Walter Dornberger (Dr. Ing. since 1935) were assigned by the Army Weapons Office to the experimental unit of Rudolf Nebel and Klaus Riedel. The basic ideas had been written down by Hermann Oberth as early as 1925. Wernher von Braun brought imaginative initiative, awareness of problems and capability to the leadership of a team, Dornberger contributed systematic procedure, organizational strength and official support, though to a modest degree.

Since 1919 there was widespread interest in rockets in many lands, inspired by technical developments in the just-ended World War. So it was only natural that Germany tried to find in this realm a way around the weapon limitations of the Treaty of Versailles. As of October 1, 1932, von Braun was therefore employed by the Army Weapons Office. When Walter Riedel, formerly at the Heylandt Works in Berlin and experienced with small rocket engines, and Heinrich Grünow, an experienced and practical worker, joined the group shortly thereafter, the first launch experiment took place at the Kummersdorf firing range south of Berlin on December 21, 1932, ending with an explosion. The next "oven", fifteen times bigger with 300 kilograms of thrust, was made of aluminum and alloys but fueled as before with alcohol and liquid oxygen. This engine, better cooled, provided 16 seconds of thrust for the first rocket, "Aggregate 1" (A 1), with its 150-kilogram weight, 1.40-meter length

and 30.4-centimeter diameter, but what with its unfinished stabilizing and nose-heaviness, still no flight. The change to high-percentage hydrogen peroxide and alcohol and fuels caused three deaths. But in Arthur Rudolph's copper powerplant they developed, on August 3, 1934, a thrust of 128 kilograms for 50 seconds. After the stabilizers were moved to the middle, between the oxygen and alcohol containers, the two "A2" test models were launched successfully at the end of December 1934 at Borkum, attaining an altitude of 2.2 kilometers. A start had been made.

The necessary establishment of the Army Test Center at Peenemünde (HVP), with its launching course from the Greifswald Oie along the coast of Pomerania, the hasty transferral of the project to there and the expansion of the HVP along with the Luftwaffe took two years!

The A3, a "pure research device", provided a thrust of 1.5 tons for 45 seconds, had stabilizing fins for the first time as well as a gas-jet rudder, thermo- and barographs and a parachute. It was 7.60 meters long and weighed 740 kilograms. By now Dornberg's group had more than 50 members, a noteworthy growth since the move in May of 1937. As of late November 1937 the rocket could be tested at the Greifswald Oie. After Dr. Walter Thiel had developed the 25-ton powerplant steadily and Dipl. Ing. Pöhlmann had improved the cooling decisively, the most important requirements for the large rockets were in existence, even though the practical development of the engine took

four more years. Four test flights of the A3 indicated a too-great weakness of the steering system against even mild side winds, despite a launch that was problem-free in and of itself.

Dornberger made sure that the next stage of development was begun only when the previous results had been checked and verified. In this way the work had to lead to success. Himself a scientist and a planner aware of his goals, he constantly convinced his team and his superiors that the difficulties would be overcome and the work would not be done by half-measures. His human relations with his colleagues and their confidence were the basis of the years-long duration of the project.

Although the design of the A4, the planned long-distance combat rocket, was in existence since 1936—it was 14 meters long, had a burning time of 65 seconds, a span of 250 kilometers and a 1000-kilogram load of explosive—four years' experience still did not solve its problems, particularly that of breaking the sound barrier. Therefore the A5 was introduced as a test model, using the motor of the A3 but improved steering, while smaller models were launched from the air to test their course stability. Test launchings in the summer of 1938 were encouraging, and after the installation of the Walter engine, three problem-free launches of the A5 took place in October of 1939, with a maximum altitude of eight kilometers; thus it remained the standard test model until 1942. An important intermediate stage had been reached! Then Hitler reduced the funds, as he only wanted to support projects that would come to fruition in "the war which will soon end." Intermediate success did not impress him; the date of mass production of the untested A4

The Beginning is Difficult

Left: The first writing of the seventeen-year-old Wernher von Braun, "Theory of Long-Range Rockets", 1929.

Below: "Minimum Rocket", called Mirak I, developed by Rudolf Nebel, Wernher von Braun and Klaus Riedel and first demonstrated publicly in September of 1931 at the "Rocket Airfield Berlin-Reinickendorf." (M)

was pushed up to May 1941. In exchange, Generaloberst von Brauchitsch made technical personnel, chosen from among 4000 men, available as of the beginning of the war.

On March 21, 1940 the 25-ton powerplant, which had been developed for four years, first ran satisfactorily for 60 seconds. Thus the main requirement for the construction of the long-range combat rocket was attained at last. Just four months later the design for the first two-stage intercontinental rocket was finished, the subsequent A9/10, which would need at least three years' development. Its maximum altitude was to be 350 kilometers, its range in 35 minutes more than 5000 kilometers. From the Atlantic coast it would reach America. The A4 powerplants were in units of six. At the same time, the Peenemünde crew tried to reduce the too-fast impact speed of the A4 in its descent by using wings with aerodynamic lift and gain range by modifying the rocket's course. The result was the winged A9, later called A4 b. A6 was never built, and A7 was the A5 with wings.

After losing the aerial battle over Britain, Hitler again recognized the urgency of the A4 program, but only within the already proceeding developments and without further funding. A visit from Armament Minister Dr. Todt with Generals Olbricht and Leeb led to no decision. Only on August 20, 1941 was Hitler moved to approve the A4 program to the point of readiness for service. When Albert Speer became Minister on February 8, 1942, he promoted the project vigorously, especially when the first test model of the A4 was ready at Peenemünde on February 25, 1942 to begin a year of testing. The launch attempts on March 18, April 29 and August 16 failed, but a thorough investigation of the failures led to constant improvements in

Rudolph's copper powerplant of 1934, an important first step.(M)

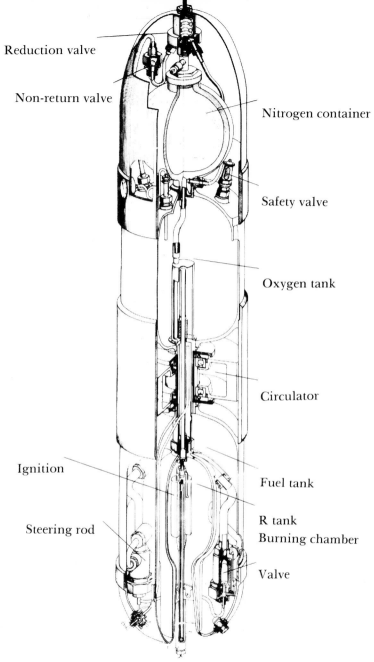

Reduction valve

Non-return valve

Nitrogen container

Safety valve

Oxygen tank

Circulator

Ignition

Fuel tank

Steering rod

R tank
Burning chamber

Valve

Diagram of the Aggregate 2, 1934.

performance. On October 3, 1942 the breakthrough success was attained: The 13.5-ton rocket raced at 1340 meters per second to a maximum altitude of 84.5 kilometers and landed, after a flight time of 296 seconds, in the Baltic Sea off the coast of Pomerania, though its steering did not function perfectly and it landed 17.7 kilometers off course. Its usefulness was proved, but improvements were still necessary. Hitler remained unmoved and demanded at least 5000 rockets in readiness for a mass attack. The Model 5 was successful, achieving a range of 147 kilometers on October 21. Models 6 and 7 were failures. At Dornberger's urging, Speer obtained Hitler's authorization for series production on December 22, 1942, though without the top level of urgency. The ban on production had expired and mass production was carefully introduced.

The series of further tests went on until September of 1943: Superb flights of up to 287 kilometers alternated with operational failures, steering malfunctions, tail fires, explosions and rockets collapsing on the launching pad or even, as on July 30, 1943, landing on the airfield of the nearby Luftwaffe test center with a thundering explosion. At the same time Speer, on his own initiative, allowed the construction of a launching bunker near St, Omer to begin. The 'Development Commission for Long-Range Firing" founded in February of that year tried to simplify the very complicated A4 so that it, with its approximately 20,000 individual parts, could be produced and assembled on assembly lines. Improvements could still be added at any time. Dornberger kept pushing for the decisive higher degree of priority, and on February 19, Hitler spoke for the first time of an "unknown, unique weapon on the way to the front." In comparison launching of A4 and Fi 103 (V 1) on May 26, two ideal launches out of 20 tests achieved a range of over 250 kilometers. From June 28 to 30, Himmler turned up at Peenemünde and took an interest in the project. The A4 cost only a third as much as a fighter plane, the Fi 103 only one forty-fifth as much as the A4. After reports from Dornberger and von Braun, Hitler finally authorized the highest level of priority on July 7. He wanted the rockets fired from a fixed launch site on the Atlantic, while Dornberger advocated mobile use in the field.

While danger from air raids increased and transfer to underground quarters in the Harz was prepared for, the command came on July 26 to set up the first units of Artillery rocket troops at Homeland Artillery Park 11 long before the rockets were ready for use at the front. A battery had three launching pads with two rockets each, an intelligence unit, a remote-control unit, an army Flak unit, a fuel supply train and firefighting troop. Their time to be ready to fire was to be 90 to 120 minutes, much superior to use at a fixed site because of their mobility. The Artillery Units (motorized) 482, 485 and 836, the SS Artillery Unit (motorized) 500 and Technical Artillery Units (motorized) 91 and 953 were formed, with twelve batteries, under Artillery Command 191/LXV. A.K. z.b.V.

Since April of 1943 the British had definite knowledge of Peenemünde, and on August 17-18 they made a mass attack on the HVP. Four weeks later, development and production recommenced, but the transfers to Blizna and the Harz were hurried; full production began there eight months later, under SS command since August 22. The targeting success was improved to 80%, and the dispersion was kept at 2% of the range. The rockets were also insensitive to weather conditions and could not be either tracked or shot down; and they carried payloads of one ton. As of May 1944 production took place at the Mittelwerk near Nordhausen.

When fire on England with the Fi 103 (V 1) ended, Paris was fired on in the night of September 5 before the second rocket attack on London from The Hague began on September 8, with four rockets launched each day in September, eight in October, twelve in November and fourteen in December, not a day going by without launches, and the number sometimes reaching 29 to 33. In all, 1115 were launched at London, 3185 at southern England and 2100 at Belgium. Eisenhower wrote: "If the Germans had succeeded six months earlier, our invasion would have proved to be difficult, if not impossible!" Dornberger, on the other hand, said in 1963: "The task was a first step to one of mankind's greatest dreams . . . a decisive turning point in human life." The pressure of circumstances at that time, though, led to a very different result!

Entwurf.

Referat VII
Aktz. 67 b 23 Wa Prw 1/VII.
Bb.Nr.0   /35g.Z.

Berlin, den 23.November 1935.

Dr.v.Braun

An
Abteilung 1. **Geheim**

Vorgang: Vortragsnotizen
Entwicklung Rauchspurgerät II
1935.

P r o g r a m m .

Auf Grund der Borkumer Ergebnisse Neukonstruktion

und Entwicklung eines Aggregats für 1500 kg Rückstoß,

45 Brennsekunden mit Steuereinrichtung nach Boykow.

Entwicklung 1.1.35 – 1.12.35.

1.) Entwicklung eines Druckzusatzsystems mit Flüssig-Stick-
stoff und Erhitzungsvorrichtung.

2.) Versuche mit Elektron-Öfen.

3.) Erprobung des 1500 kg Ofens.

4.) Entwicklung einer Einebenden-Steuermaschine und Durcher-
probung (zweite Einebende-Steuermaschine im Bau).

5.) Haltbarkeitsversuche mit hochhitzebeständigem Gasstrahl-
rudern.

6.) Festigkeitstechnische Untersuchungen und fertigungstech-
nische Entwicklung an Elektron-Behältern für Innen- und
Außendruck.

7.) Beginn der Entwicklung von Kreiselpumpen für Brennstoff
und Flüssig-Sauerstoff in Zusammenarbeit mit Klein,
Schanzlin & Becker.

8.) Beginn der Entwicklung eines Rauchspur-Jagdflugzeuges in
Zusammenarbeit mit L.C.II, Junkers und Heinkel.

9.) Entwicklung einer Rauchspur-Starthilfe für überladene
Bomber in Zusammenarbeit mit L.C.I und Versuchsanstalt
für Luftfahrt.

10.) Bauliche Erweiterungen und Begründung der "Versuchsstelle
West".

Auf Anordnung des Gruppenleiters
1 Abschrift für Wa Prw 8/V angefertigt.

Zu den Akten

so ist das tatsächliche Profil

Sollprofil der Düse

ca. 0,5 mm

Schematische
Skizze

A document concerning subsequent project direction, with the signature of Wernher von Braun.

Wernher von Braun's freehand drawing of an improved shape of the jet exhaust duct—care given to the smallest detail.

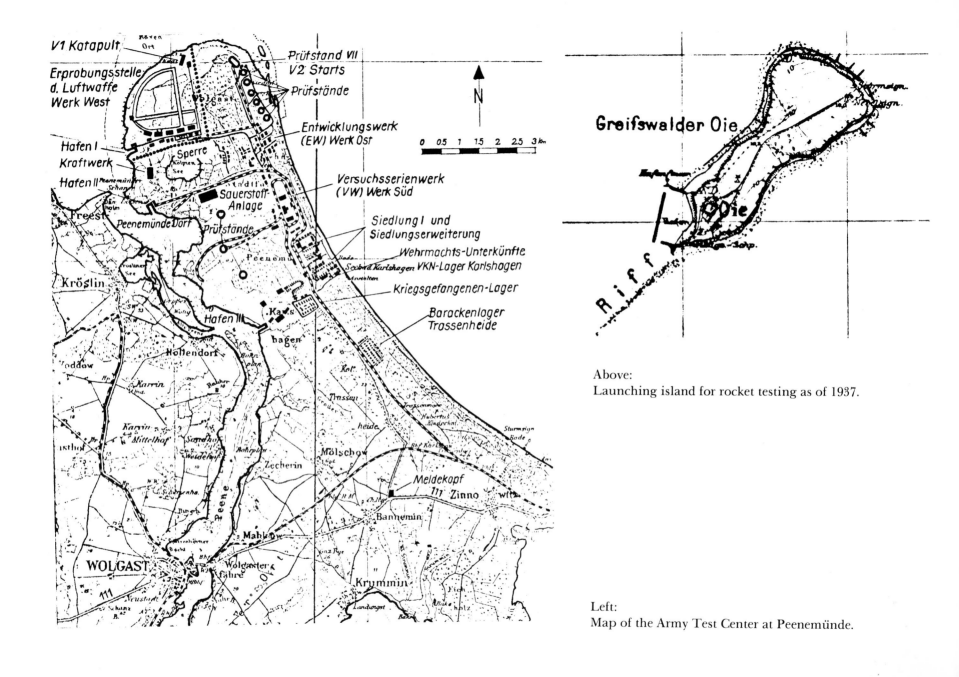

V1 Katapult

Erprobungsstelle
d. Luftwaffe
Werk West

Hafen I

Kraftwerk

Hafen II

Freest

Peenemünde Dorf   Prüfstände

Kröslin

Hafen III   Ka..s

..hagen

Hollendorf

..oddow

Karrin

Karrin
Mittelhof   Sandhof

..sthof

Zecherin

Mölschow

WOLGAST   Wolgaster
Fähre

111   Neustadt

Krummin

Sperre
Kolpien
See

Sauerstoff-
Anlage

Peenem..

Meldekopf
111 Zinno..   ..itz

Bannemin

Mahlzow

Prüfstand VII
V2 Starts

Prüfstände

Entwicklungswerk
(EW) Werk Ost

Versuchsserienwerk
(VW) Werk Süd

Siedlung I und
Siedlungserweiterung

Wehrmachts-Unterkünfte
Seebad Karlshagen VKN-Lager Karlshagen

Kriegsgefangenen-Lager

Barackenlager
Trassenheide

N

0 0.5 1 1.5 2 2.5 3 km

Greifswalder Oie

Riff

Above:
Launching island for rocket testing as of 1937.

Left:
Map of the Army Test Center at Peenemünde.

The subsequent Peenemünde firing range along the Pomeranian coast, with tracking stations on land.

Erecting the A3 with the new erecting apparatus at the Greifswald Oie, early December 1937.(M)

The stamp says "State Secret!"

The stamp says "State Secret!"

Staatsgeheimnis!

Geheimhaltungsverpflichtung beachten!

Dies ist ein geheimer Gegenstand im Sinne des § 88 Reichs... ...etzbuchs (Fassung v. 24. April 1934). ... ...ch wird nach den Bestimmungen dieses G... ...s bestraft, sofern nicht andere Strafbestimmungen in Frage kommen.

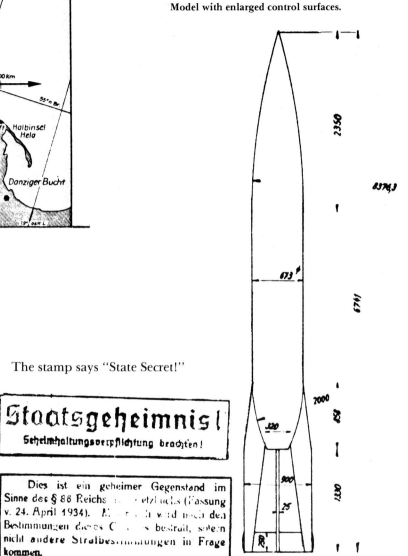

Modell mit vergrößertem Leitwerk

Model with enlarged control surfaces.

Construction drawing of the A3 with improved control surfaces for the models of the supplying firms.

Above: The powerplant of the second A3 test model is recovered from the Baltic Sea on December 6, 1937.

Above: The A3 rocket stands ready for a series of tests from December 4 to 11, 1937.(M)

Right: Oxygen transport on the field railway lines to fuel the A3.(M)

A3 rockets with their scaffolding on the Oie in mid-December 1937, covered with tarpaulins as weather protection, before launching.(M)

First results are shown

A 5   ## A 3

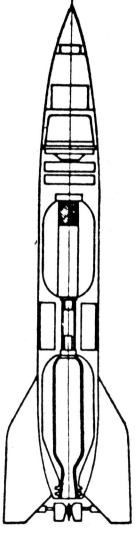

From A3 to A5: extended control surfaces, shorter combustion chamber now outside the alcohol tank—the design has matured.

Above: A Heinkel He111 with A5 rocket models ready to launch from high altitudes to improve course stability with the help of built-in instruments, subsequently recovered by parachute, very successful tests by Dr. Ing. Steinhoff.(M)

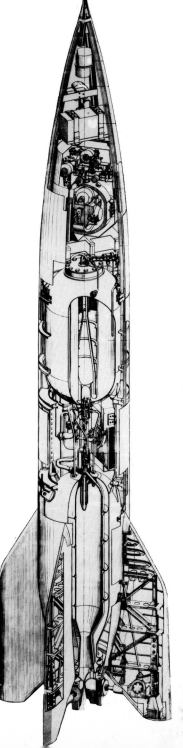

Right: Interior design drawing of the A5, semi-schematic.(M)

Wernher von Braun criticizes a report by Engineer Groth on tests of the A4.

AUSZUG aus dem BERICHT des.Ing.GROTH über VORVERSUCHE mit A 4
(66/29 g.Kdos.)

Das Ergebnis dieser Untersuchung
ist,daß mathematische Stabili-
tätsuntersuchungen (sei es durch Bahn-
rechnungen oder durch die Methode der kleinen Schwingungen)
im Falle des ungesteuerten R-Kör-
pers Stabilität liefern, während
in der Wirklichkeit angefachte
Taumelschwingungen auftreten.

Hiernach müssen auch die Ergebnisse am gesteuerten R-Körper
als nicht völlig gesichert ange-
sehen werden, da sie unter den
gleichen Voraussetzungen und
mit den gleichen Methoden gewon-
nen sind.

A view of the open rear end of an A4: at upper left is the hydrogen peroxide tank for the steam system, below the injector of the powerplant with 18 circular jets and mixing chambers, as used in 1941-42, later simplified for large-series production.(M)

Heizbehälter: Betriebszustand
(theor. Werte)

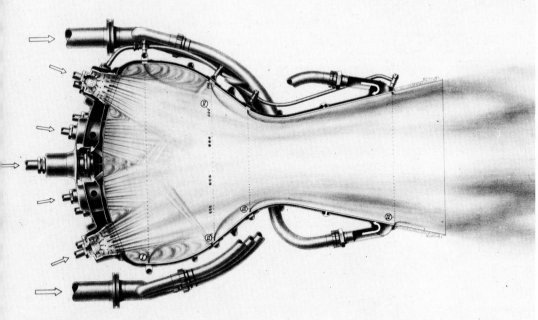

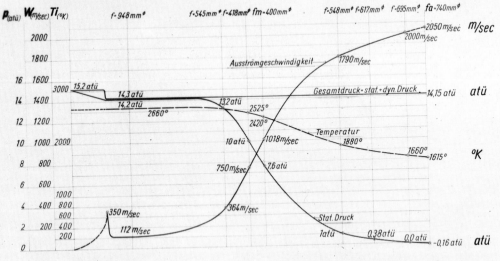

Brennstand ähnlich Prüfstand VI, besondere Räume inner-
halb des Brennstandes unter der Erde für N_2-Batterien,
Auroltank und Minimax-Löschanlagen.

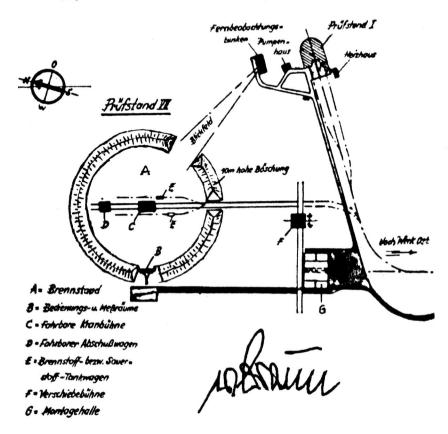

A = Brennstand
B = Bedienungs- u. Meßräume
C = Fahrbare Kranbühne
D = Fahrbarer Abschußwagen
E = Brennstoff- bezw. Sauer-
     stoff-Tankwagen
F = Verschiebebühne
G = Montagehalle

Above: Drawing of Test Site VII on the northwest
edge of the Peenemünde-East development facility,
from which most A4 rockets were launched.

Left: The heat retainer of the A4 with the physical
data in operating condition.(M)

Right: Distinguished visitors at Peenemünde in 1941: from left to right, General of the Artillery Emil Leeb, Chief of the Army Weapons Office. in the background Heinrich Lübke, managing director of a construction firm active in Peenemünde, Armament Minister Generalmajor Dr. Todt, Oberst Dornberger, General of the Infantry Friedrich Olbricht, OKH, Chief of the General Army Office.(M)

Soldiers and Technicians develop the space rocket.

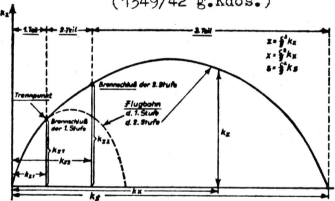

Flugbahn der Zweistufen-Rakete
(1349/42 g.Kdos.)

Above: As early as 1942 the intercontinental two-stage rocket had been designed.

Left: Powerplant specialist Dr. Thiel gives information on Model 3 of the A4 rocket in July 1942 to his team, from left to right, Hunter, Hainisch, Schwarz, Muenz, Dr. Hackh, Zoike, Dr. Schilling.(M)

Below: In the summer of 1942 the first underwater launches of 21-cm solid-fuel rockets from a U-Boat of the IX C-Class were made, the tests being satisfactory for subsequent rocket launches.(M)

Preparations for the launch of the same model, launched on August 16, 1942 with useful success— the witch is meant to symbolize space travel and the ''magic of technology.''(M)

Behind the ten-meter protective wall of Test Site VII, an A4 with a portable gantry is being fueled from a tank truck.(M)

The A4 is successful

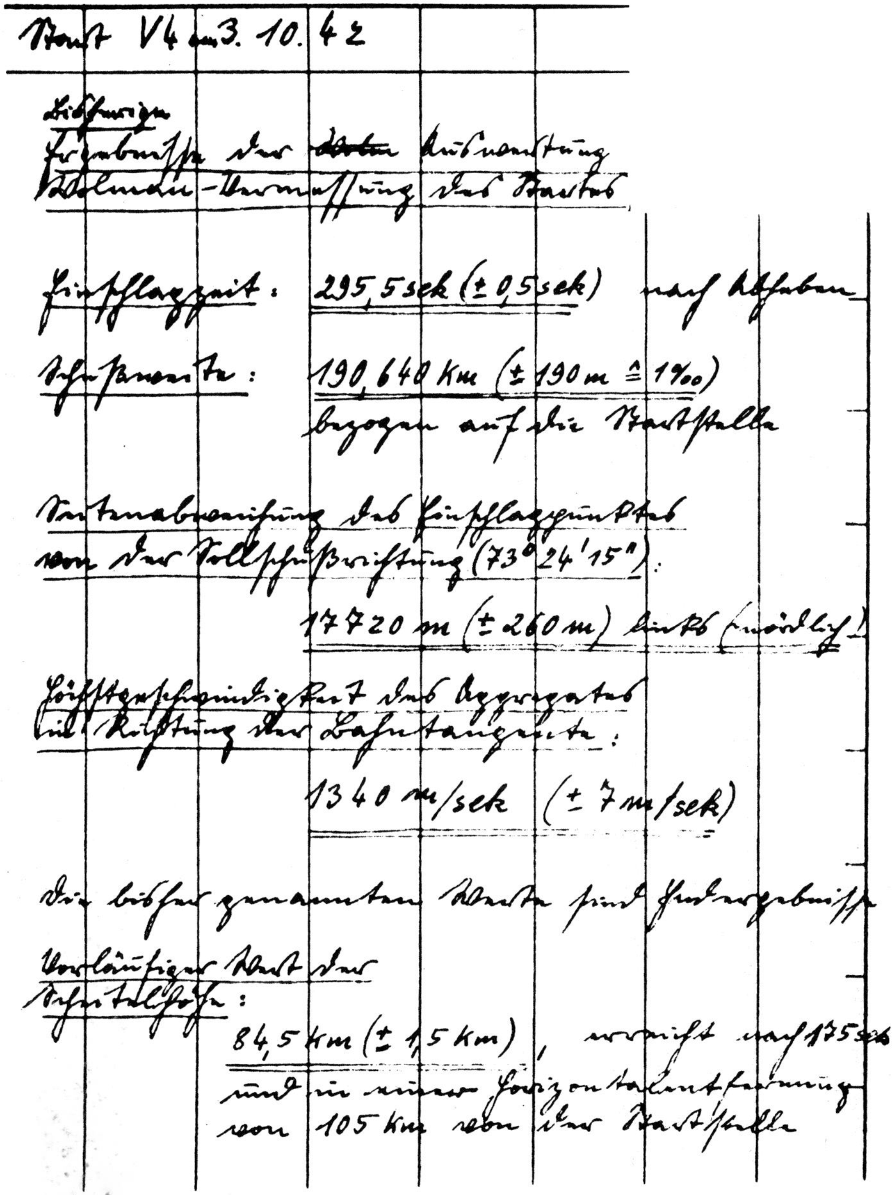

Start V4 am 3. 10. 42

[handwritten log entries in German cursive]

Feinflugzeit: 295,5 sek (± 0,5 sek) nach Abheben

Schussweite: 190,640 Km (± 190 m ≙ 1‰)
bezogen auf die Startstelle

Seitenabweichung des Feinflugpunktes
von der Sollschussrichtung (73° 24' 15"):

17720 m (± 260 m) links (nördlich)

Höchstgeschwindigkeit des Aggregates
bei Erreichung der Bahntangente:

1340 m/sek (± 7 m/sek)

Die bisher genannten Werte sind Fundwergebnisse

Vorläufiger Wert der
Scheitelhöhe:

84,5 Km (± 1,5 Km), erreicht nach 175 sek
und in einer Horizontalentfernung
von 105 Km von der Startstelle

Above: Lower part of the testing tower for the A4 at
Test Site VII in Peenenünde in 1942.(M)

Left: Original record of the pioneering launching
success of the A4 at Peenemünde on October 3, 1942.

Right: Atomic power for rockets was considered as early as 1942, as this high-priority secret research report by Colonel Janssen of HVP to a cover address shows; it was discovered 27 years later by the U.S. Project "NERVA."

Below: When Model 9 of the A4 left its course on December 12, 1942 and was destroyed on impact, the recovered injector head of the powerplant was carefully inspected for faults.(M)

9 Ausfertigungen
Abdr.I.Rechnungslegung (3.Ausfertigung)

Oberkommando des Heeres
(Chef der Heeresrüstung und
Befehlshaber des Ersatzheeres)

Berlin W 35, den 15.10. 1942
Tirpitzufer 72—76
Fernsprecher:

Auftrag-Nr.: Wa Prüf 11 KAP/L SS 011-5371/42 EY 200

Auftrag-Nr. bei allem Schriftwechsel stets angeben!

11.11.1944

Bb-Nr. 959/42 gKdos Geheime

Kriegsauftrag

Firma **Forschungsanstalt der Deutschen Reichspost**
z.Hd.v.Herrn Postrat K u b i c k i

W-Nr

Berlin - Tempelhof

Ringbahnstr.125

Es wird Ihnen hiermit der Auftrag übertragen auf

1.) Durchführung grundsätzlicher Untersuchungen über die Leistungssteigerung von Flüssigkeits-R-Antrieben durch Verwendung von Treibstoffgemischen höchsten Energiegehaltes.

2.) Untersuchung der Möglichkeit der Ausnutzung des Atomzerfalls und Kettenreaktion zum R-Antrieb.

Der Preis versteht sich: — — — —

Zahlungsbedingungen: — — — —

Dem Auftrag liegen die "jenigen — in Ihren Händen befindlichen — "Besonderen Bedingungen für Kriegsaufträge des Heereswaffenamts (Ausgabe vom März 1941)" und die weiteren hierin genannten Unterlagen zugrunde.

How its use in the west was planned

Below: Design for a fixed battery under concrete on the Atlantic coast, as Hitler wanted.

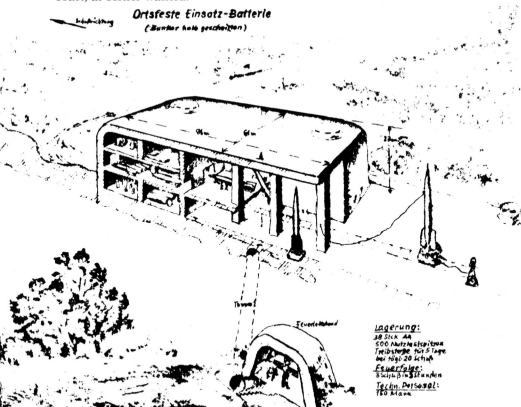

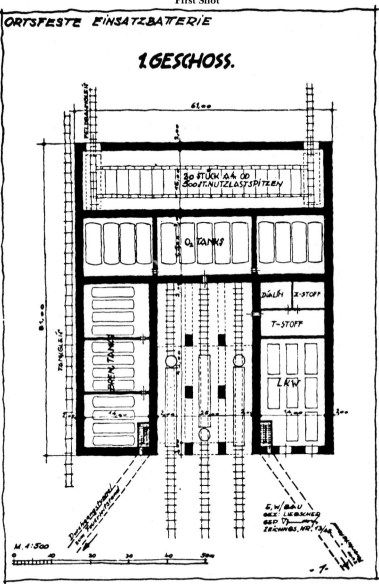

Firing plan of the bunker near Wizernes in 1942. The rocket hall extended through several stories.

The long course of testing is accompanied by many setbacks

Left page: At the launch of the tenth model of A4 at Test Site VII in Peenemünde at 10:45 A.M. on January 7, 1943, 2.5 seconds after ignition of the first stage there was an explosion in the combustion chamber on account of a malfunction.(M)

Above: For the transport and erecting of the A4, Meiller trailers with typical clamps for the collars were used.(M)

Left: Afterward, the rocket fell burning to the side before power was turned on for the main stage. In seconds a sea of flames flared up.(M)

Left: Walter R. Dornberger, born in Giessen in 1895, entered service in 1914, became Doctor of Engineering on March 5, 1935, Colonel on August 1, 1940, seen here as Major General (since June 1, 1943), taken prisoner by the British and sent to the USA and to Wright-Patterson Air Force Base in Dayton, Ohio, became Vice-President of the Bell Aero-System Company in 1960 and died near Frankfurt on June 28, 1980.

Organizer, Theoretician and Work

Below: Major General Dr. Dornberger talking with Professor Hermann Oberth, the pioneer of space travel theory, in the summer of 1943.

The successful launch of an A4 from Test Site VII at Peenemünde in 1943.(M)

On July 6, 1943 the Model 33 of the A4 was destroyed at Test Site VII by premature cessation of burning.(M)

Setbacks
again

Below: Every recovered piece of an exploded rocket is thoroughly analyzed for faults from which lessons can be learned.(M)

Above: The same rocket tips over and burns out.(M)

Left: Successful liftoff of an A4 four seconds after launching in the summer of 1943, seen from Test Site I.(M)

The development of rocket artillery begins

Below: Inspection by the Army Weapons Office in 1943: from left to right Major General Dr. Dornberger (seen from rear), in civilian clothes Dr. Herrmann, Director of the supersonic wind tunnel, Lieutenant General Schneider, Chief of the Army Weapons Office, with binoculars, in front Dr. von Braun, Colonel Zanssen, Commander of Peenemünde.(M)

Entwurf

Geheime Kommandosache

Wa A
11 e 36 Wa Prüf 11/Stab B
1091/43g.Kdos.

Berlin, den 26.Juli 19.
J 3 0931, App. 20
Oberstleutnant Thom

6 Ausfertigungen
Verteiler:

AHA / In 4	= 1.Ausf.
Wa A / Stab Chefgr.	2.Ausf.
Wa Prüf/Stab In	= 3.Ausf.
KAP	= 4.Ausf.
Wa Prüf 11/Stab B	
f. 72 p 66	= 5.Ausf.
Entwurf	= 6.Ausf.

An
ARA/In 4

Betr.: Sonderformationen.

Vorg.: Besprechung am 21.7.43 mit Oberstleutnant Thom (Wa Prüf 11).

I. Für die Aufstellung der Sonderformationen ist folgender 1.Aufstellungsplan vorgesehen:

1.Aufstellungstag:	Truppenteil:	Aufstellungsort:	Vorläufige Verwendung
15.8.	3.Battr.(?) Art.Abt. (t mot) 953	Carlshagen	Ausbildung. – ab 1.9. mit Teilen zum Einsatz.
1.9.	Stab Art.Abt. (t mot) 953	Greifswald	Ausbildung.
1.9.	Stab Art.Abt. (mot) 836	Anklam	Ausbildung. – ab 1.10. nach Gr. Born.
15.9.	Stab Art. Kdr. (mot) 191	Carlshagen	Ausbildung.
15.9.	1., 2., 3. Battr. Art.Abt. (mot) 836 mit 1.-3.Treibstoffkolonne	Gr. Born	Ausbildung bis zum Einsatz
15.9.	1.u.2. Battr. Art.Abt. (t mot) 953	Carlshagen	

Wa A – Wa Prüf 11 – übersendet in den nächsten Tagen die von AHA/In 4 mitgeprüften K St N. Aus Geheimhaltungsgründen wird geboten, keinen Gesamtaufstellungsbefehl herauszugeben, sondern Einzelbefehle.

II. Stab AR 760 wird zu einem noch festzusetzenden Zeitpunkt in Stab Art.Abt. (mot) 836 überführt.

Secret plan for the establishment of rocket artillery troops, July 26, 1943.

Preparations for use

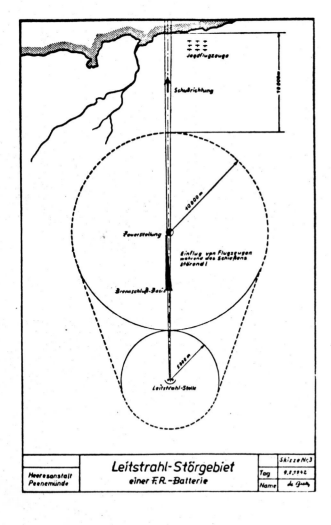

From top to bottom:
Warhead with 1000 kg payload
Navigational devices
Gyroscope
Alcohol tank
Automatic valve
Oxygen tank
Permanganate tank
Double-walled alcohol ducting
Tank for hydrogen peroxide
Steam turbine
Pumps
Oxygen distributor
Combustion chamber
4 Graphite rudders
Jet
Drawing 71/72
AGGREGATE 4
V-2 on the Meiller truck (V-2)

Left:
Radio-control drawing of a rocket battery.

Right:
The Meiller trailer serves in practice as a "mount" for the A4.(BA)

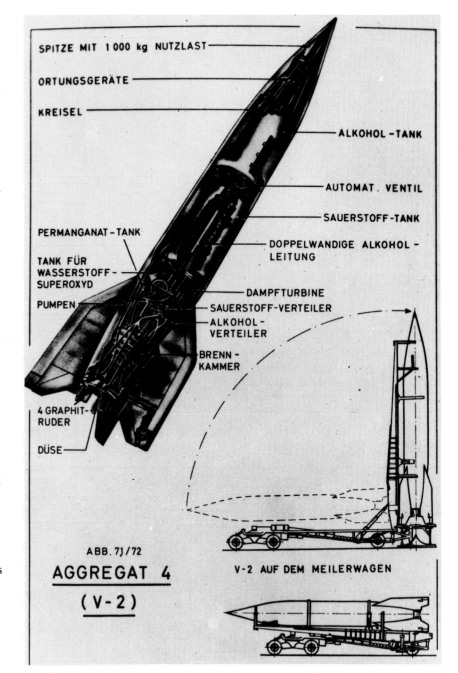

25

Tower truck of an **FR** battery with hydraulic platform for further servicing of the vertically erected rocket.(M)

Air attack in August of 1943

Below: The fully destroyed main street of the scientific community of Karlshagen after the large-scale air raid on August 17-18, 1943.(M)

Above: The air raid of August 17-18, 1943 caused only limited damage in the workshops of the HVP.(M)

View of the destroyed Karlshagen community after August 17-18, 1943.(M)

Transfer of firing experiments to the east

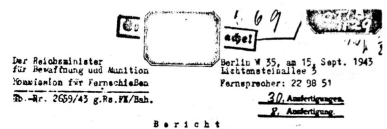

Berlin W 35, am 15. Sept. 1943
Lichtensteinallee 3
Fernsprecher: 22 98 51

30. Ausfertigungen
2. Ausfertigung.

B e r i c h t
über die Sitzung der Kommission für Fernschießen am 9.9.1943

- -

Im Laufe der Sitzung wurden für das Gerät A 4 folgende Termine festge-
stellt:

    1) Erprobung des neuen Ofens am 15.9.1943
    2) Erster vereinfachter scharfer Schuß Mitte Oktober 1943
    3) Erster gezielter scharfer Schuß am 15.11.1943
    4) Erste mot.Batterie steht am 1.12.1943
    5) Nächste Kommissionssitzung Anfang Oktober 1943.

Zu Punkt 1) der Tagesordnung:

Prof.v.Braun berichtet kurz über den Stand der Entwicklung A 4 und an-
         schließend kurz über Wasserfall.

Die Entwicklung des Gerätes A 4 ist praktisch zum Abschluß ge
kommen. Schießversuche mit scharfer Sprengladung (dynamische
Versuche) stehen noch aus. Diese müssen noch stattfinden,
um sich ein Bild über die Trefferwirkung und über die Funk-
tion des Zünders zu machen. Statische Versuche sind schon
gemacht worden, indem Geräte gesprengt wurden. Die Wirkung
entsprach der Wirkung einer Mine von etwa 1300 kg. wobei ein
Krater von 7 m Tiefe mit einem Durchmesser von 13 m ent-
stand. Über die kombinierte Wirkung des im Ziel auftreffen-
den Geschosses sind noch keine Erfahrungen vorhanden.

Auf Befehl des Führers wird nun die Anlage ganz nach dem
Osten verlegt, um dort ungestört von Luftangriffen Schießversuche durchführen zu können. Die erste mot.Batterie
steht erst am 1.12.1943 zur Verfügung. Die Gründe hierfür
liegen darin, daß fertiggestellte Schaltanlagen und viele
Einzelteile bei dem letzten Luftangriff in Peenemünde zer-
stört worden sind. In der letzten Zeit haben wir uns tot ge
stellt, wir müssen jedoch einen kurzen Schießbetrieb wieder
aufleben lassen, doch werden diese Versuche nur bei schlech-
tem

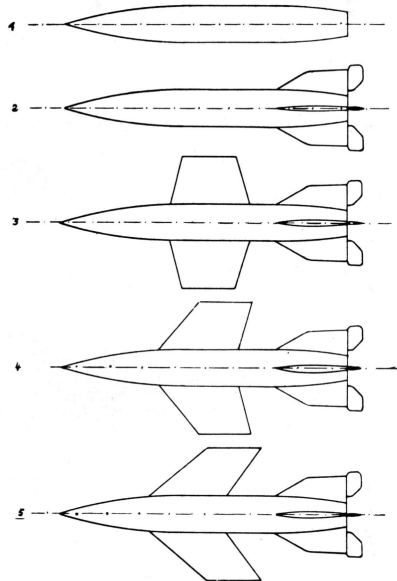

Above:
The models for the A4b; the decision is made in favor of Model 5, to utilize its ability to glide.

Left:
Plans for the "Commission for Long-Range Firing" of September 15, 1943 for continuation of training at the "Heath Camp" of Blizna/Gen. Gouvern.

An A4 in a successful launch in the autumn of 1943 at Blizna, spied on for a long period by the Polish Resistance.

Much preparation in setting up rocket batteries

Right: After the rockets are erected, checking the steering system of the nearest one has already begun.(M)

A tank truck of the fuel supply train in service with the 1st and 2nd units.

Right: An A4 at the "Central Works" in Niedersachswerfen after the Americans arrived on July 3, 1945, two days before the Russians.

Below: A fuel supply train's tank trailer for liquid oxygen.

Production continues underground, especially in the Harz.

Towing an unfueled A4 rocket out of an isolated forest depot by a narrow-gauge locomotive.(M)

Left: Delivering a rocket to the preparation facility on a Vidal transporter.(BA)

Typical transport trailers, slightly modified, serve to transport the big rockets.

Below: The Meiller trailer brings it to the firing position and erects it.

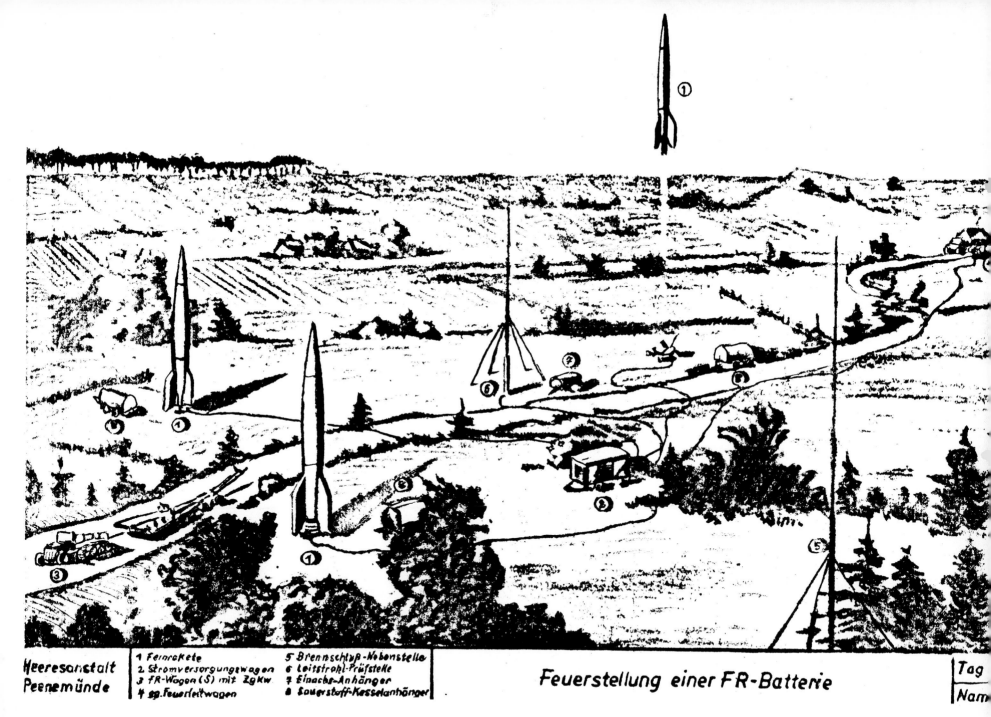

1 Fernrakete
2 Stromversorgungswagen
3 FR-Wagen (S) mit ZgKw
4 sp. Feuerleitwagen

5 Brennschluß-Nebenstelle
6 Leitstrahl-Prüfstelle
7 Einachs-Anhänger
8 Sauerstoff-Kesselanhänger

Feuerstellung einer FR-Batterie

Tag
Nam

Drawing of a long-range rocket launching site after being set up and firing its first rocket.

Erection of the A4 rocket by a Meiller trailer; the towing tractor has already left. Model for later simple setups.(M)

Right: RAF photo of November 27, 1944. showing the forward portion of the A4 powerplant with its 18 jets and parts of the drive block, discovered in Belgium—the ruins reveal the construction.(E)

The A4 is in mobile use as "V2", and successful at great expense

Left: Launch of an A4, now called V2, from a field launch near Wassenaar, The Netherlands.(BA)

Assembly hall at the HVP Test Facility South in 1944, with A4 rocket powerplants; more than 60,000 changes were needed to prepare them for series production—the powerplant determined the quality of the rocket.(M)

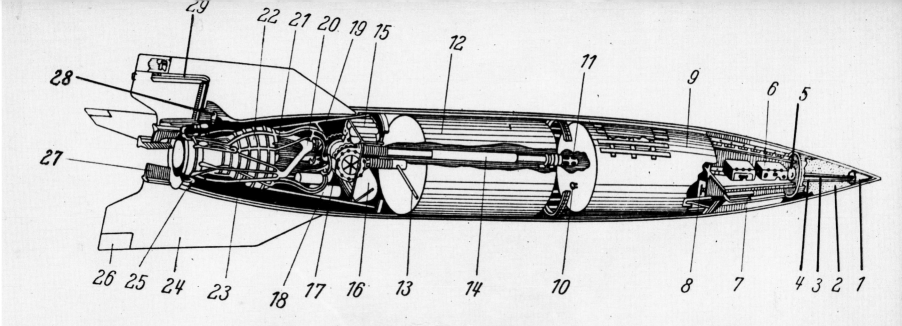

V 2

1. Schlagbolzen
2. Sprengladung
3. Schlagröhre
4. Elektrisches Kabel
5. Elektrischer Zünder
6. Funkgerät (FUMG)
7. Stickstoffbehälter
8. Autom. Steuerapparat
9. Methylalkoholbehälter
10. Absicherung für Behälter 9
11. Absperrventil
12. Behälter für flüssige Luft
13. Absperrventil für Behälter 12
14. Zuleitungsrohr
15. Preßluftflaschen

16. Zwischenbehälter für Sauerstoff
17. Turbinen und Pumpen
18. Behälter für Permanganat
19. Sauerstoffverteilerleitung
20. Alkoholverteilerleitung
21. Einspritzdüsen
22. Verbrennungskammer
23. Kühlleitungen
24. Stabilisierungsflächen
25. Rudermotor
26. Äußere Ruder
27. Innere Ruder
28. Elektromotore
29. Steuergestänge

Rauchen verboten.

Above:
The A4 as a railroad rocket in transport through thick woods.(M)

Eight A4 rockets, camouflaged and examined, at Peenemünde in 1944.

Above: Rear end of a rocket on a stake truck in front of a tunnel.

A4/V2 before being launched from a railroad car, a technically interesting possible utilization that was hardly ever used but still plays a role in the USA today.

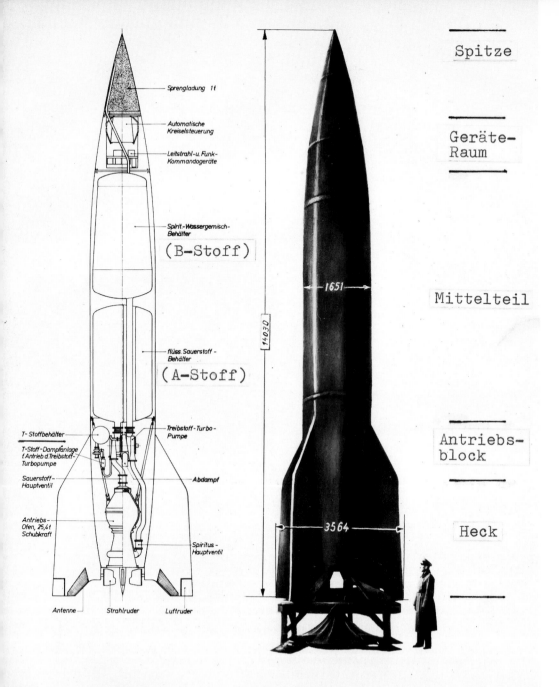

Sprengladung 1t

Automatische Kreiselsteuerung

Leitstrahl-u. Funk-Kommandogeräte

Spirit.-Wassergemisch-Behälter

(B-Stoff)

flüss.Sauerstoff-Behälter

(A-Stoff)

T-Stoffbehälter

T-Stoff-Dampfanlage f.Antrieb d.Treibstoff-Turbopumpe

Sauerstoff-Hauptventil

Antriebs-Ofen, 25,4t Schubkraft

Antenne   Strahlruder   Luftruder

Treibstoff-Turbo-Pumpe

Abdampf

Spiritus-Hauptventil

Abb.1 Fernrakete im Schnitt.

1651

14030

3564

Abb.2   Fernrakete A4

auf der Startplattform.

Spitze

Geräte-Raum

Mittelteil

Antriebs-block

Heck

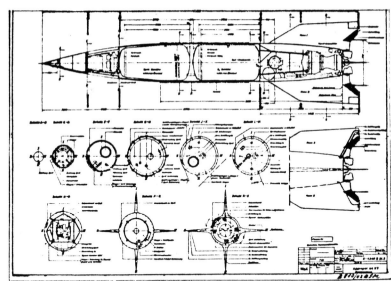

Above:
Comparison sketch and cross section of the A4/V2.

Comparison of the A4 in cross section, and on a launching pad.

Inspection in the Spring of 1944 by Generalfeldmarschall Keitel. From left: General Warlimont, OKW, Generalfeldmarschall Keitel, Generaloberst Fromm, Commander Ersatzheer, Begleitoffizier, and Generalmajor Dr. Dornberger.

The age of rockets has dawned, with futuristic designs.

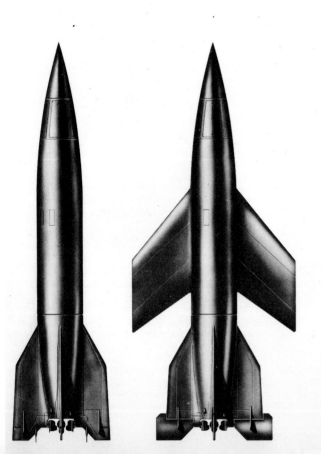
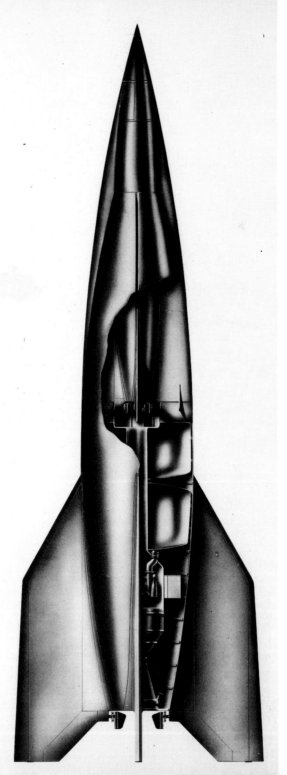

Left: **A4**

Stages: 1
Length: 14.03 meters
Weight: 12.9 tons
Thrust: 25 Mp
Load: 1 ton
Speed: 5760 kph
Range: 330 km
Development: 1936-1944
First launch: October 3, 1942
Final status: Series production

Center: **A4b**

Stages: 1
Length: 14.03 meters
Weight: 13.5 tons
Thrust: 25 Mp
Load: 1 ton
Speed: 5500 kph
Range: 750 km
Development: 1940-1945
First launch: January 24, 1945
Final status: Testing

Right: **A 9/A 10**

Stages: 2
Length: 87 meters
Weight: 200 tons
Thrust: 13 Mp
Speed: 4320 kph
Range: max. 5500 km
Development: 1940-1944
First launch: None
Final status: Pre-project

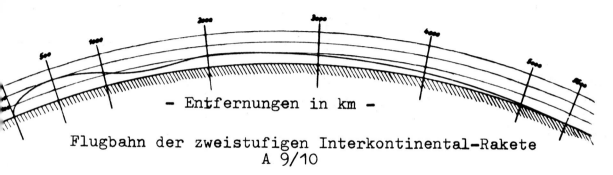

- Entfernungen in km -

Flugbahn der zweistufigen Interkontinental-Rakete
A 9/10

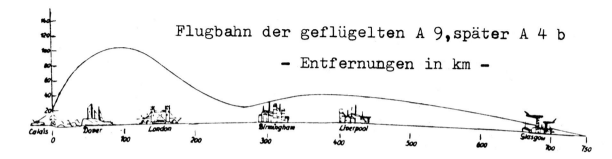

Flugbahn der geflügelten A 9, später A 4 b

- Entfernungen in km -

Above: Trajectories of the A9 and A9/10 with flattened descent as a result of gliding.

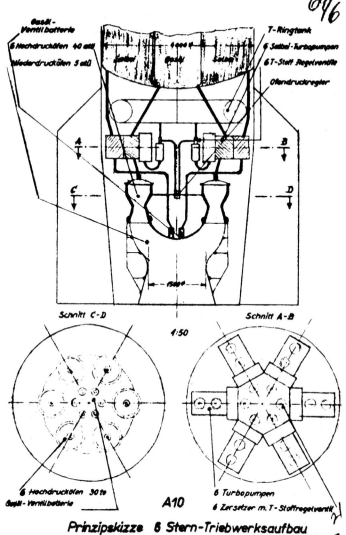

A10

Prinzipskizze 6 Stern-Triebwerksaufbau

Above: Bundles of six units in the powerplant of the A9/10, with powerful thrust.

Left: Field Marshal Keitel observes the flight of A4 rockets through a filming theodolite in 1944; at his right are Lieutenant General Beisswänger, Chief of Army Armament, Major General Dr. Dornberger and Professor Hermann Oberth (obscured).(M)

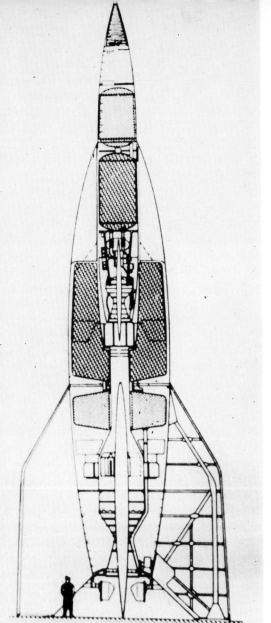

Right: Transcript of the discussion of December 9, 1944 concerning the use of the A4 from the sea, towed by U-boats.

Above: The model of the gigantic A9/10 rocket gives an impression of its great size.

Geheime Kommandosache!

Chef-Sache
Nur durch Offizier

rb. 1337   /727

2. 94/44 gKdos.Chefs

O.U.,den 11.Dezember 1944.

4 Ausfertigungen
1.Ausfertigung: General Rossmann
2.Ausfertigung: Direktor Riedel III, E.W.
3.Ausfertigung: Dr.Dickmann, Vulkan Werft
4.Ausfertigung: Entwurf, Wa Prüf(BuM) 10/I

N i e d e r s c h r i f t

über die Besprechung vom 9.12.1944 bei Wa Prüf(BuM) 10.

Teilnehmer: Wa Prüf(BuM)10: Generalmajor Rossmann   Abt.Chef
                             Oberstltn.   Börgemann   Abt.Chef z.b.V.
                             Major   Schneider   Gruppenleiter I
                             Dr.Ing.   Jauernick   für Gruppenl.II
                             Major   Wenzel   Gruppenl.III
                             Hauptmann   Hofmann   Gruppenl.IV
                             Oberinspekt. Schuchmann für Gruppenl.V

             Vulkan Werft Stettin: Dr. Dickmann

             E.W. Karlshagen: Direktor   Riedel III
                               Direktor   Hüter
                               Dipl.Ing.   Lührsen
                               Dr.   Debus

Gegenstand der Besprechung:

        Schiessen mit A4 von See aus.

Zweck der Besprechung:

        Erste technische Fühlungnahme zwischen der Werft und der
        Entwicklungsabteilung des Gerätes.Vorklärung und Fixierung
        einiger grundlegender Fragen über die schiesstechnische
        Durchführbarkeit des Vorhabens.

Inhalt der Besprechung:

Dr.Dickmann erläutert den Plan,das Gerät A4 in einem von einem U-Boot
unter Wasser geschleppten Schwimmkörper auf günstige Schussposition auf
eine feindlich Küste heranzubringen,das Gerät von dem in Schussstellung
gebrachten Schwimmkörper zu verschiessen und diesen zu neuer Verwendung
wieder zum Heimathafen zurückzuschleppen. Schiffsbaumässig ergeben sich
hier vor allem die Abmessungen,der geforderten Stabilität und weiterer,
durch die Fragen der Eigenarten des Gerätes und des Abschusses beding-
ten Einrichtungen.

Above: President John F. Kennedy at the Marshall Space Travel Center in 1962, visiting Wernher von Braun at the height of his success, seven years before the landing on the moon. The managing director of NASA since 1970, von Braun died in 1977.

On its way to being an international rocket, the successful subsequent model of the A4 was halted immediately because of the war situation.

Right: A4 b, forerunner of the A9, shortly before its successful launch on January 24, 1945 at Test Site X in Peenemünde—the ancestor of the "Space Shuttle." (BA)

„UHU"-He 219

THE BEST NIGHT FIGHTER OF WORLD WAR II

SCHIFFER MILITARY

ARMORED MILITARY VEHICLES

MAUS

AND OTHER GERMAN ARMORED PROJECTS

DORNER DO 335 "PFEIL"

THE LAST AND BEST PISTON-ENGINE FIGHTER OF THE LUFTWAFFE

GERMAN

ARMORED TRAINS

IN WORLD WAR II

SCHIFFER MILITARY · VOLUME 17

GERMAN MOTORCYCLES IN WORLD WAR II

SCHIFFER MILITARY

ALSO FROM:

•SCHIFFER MILITARY HISTORY•

•THE WAFFEN-SS•THE HG PANZER DIVISION•
•THE 1ST SS ARMORED DIVISION•
•THE 12TH SS ARMORED DIVISION•
AND MORE...

THE WORLD'S FIRST JET BOMBERS

ARADO AR 234
JUNKERS JU 287

by FRANZ KOBER

ARADO AR 234 B-2

SCHIFFER MILITARY

GERMAN AIRSHIPS

PARSEVAL-SCHÜTTE-LANZ-ZEPPELIN

HEINZ J. NOWARRA   -SCHIFFER MILITARY-

GERMAN BATTLETANKS

"NEWLY BUILT VEHICLE"-PANZER I-PANZER II-PANZER III
-PANZER IV- PANZER V-"PANTHER"-PANZER VI "TIGER"
and "KING TIGER"-"MAUS"

IN COLOR 1934-45

SCHIFFER MILITARY · VOLUME 18

GERMAN NIGHT FIGHTERS

IN WORLD WAR II

MANFRED GRIEHL

Ar 234-Do 217-Do 335-Ta 154-He 219-Ju 88-Ju388-Bf 110-Me 262 etc.

-SCHIFFER MILITARY-

The King Tiger Tank

by Horst Scheibert

THE PANTHER FAMILY

by HORST SCHEIBERT

SCHIFFER MILITARY

GERMAN PERSONNEL CARS WARTIME

"The light, medium and heavy personnel vehicles of the ..."

CAPTURED TANKS UNDER THE GERMAN FLAG

RUSSIAN BATTLE TANKS

DR. WERNER REGENBERG
HORST SCHEIBERT

-SCHIFFER MILITARY-

GERMAN   SHORT-RANGE RECONNAISSANCE PLANES

1930-1945